FOR YOUR HOME

FLOOR TREATMENTS

FOR YOUR HOME

FLOOR TREATMENTS

CANDACE ORD MANROE

Little, Brown and Company
Boston   New York   Toronto   London

DEDICATION

FOR MEAGAN AND DREW

First Edition

ISBN 0-316-65211-3

Library of Congress Catalog Card Number 93-81275

A FRIEDMAN GROUP BOOK

For Your Home: Floor Treatments
was prepared and produced by
Michael Friedman Publishing Group, Inc.
15 West 26th Street
New York, New York 10010

Editor: Sharyn Rosart
Designer: Tanya Ross-Hughes
Art Director: Jeff Batzli
Photography Editor: Jennifer Crowe McMichael

10 9 8 7 6 5 4 3 2 1

Published simultaneously in Canada by Little, Brown & Company (Canada) Limited

Printed in China

Grateful acknowledgment is given to architects, designers, and photographers. Every effort has been made to correctly credit contributors to the projects. In the case of any omissions, the publishers will be pleased to make suitable acknowledgments in future editions.

TABLE OF CONTENTS

INTRODUCTION

In home design, the floor is much more than a mere surface that enables movement from one space to the next. Akin to a fifth wall, it demands both practical and aesthetic consideration. Given that the floor spans the entire area of a room, it is one of the most important architectural surfaces in the space: Nothing carries as much potential impact.

As today's approach to interior design loosens up, becoming less structured and predictable, and more personal, homeowners everywhere are joining professional designers in a fresh willingness to experiment and take bold chances, even at the risk of major mistakes. And, to the surprise of many, even seeming mistakes usually have a way of working out, often becoming the most outstanding design coup in the space.

One of the results of this newfound confidence in decorating is that the options in flooring treatments have increased at a burgeoning pace. The old standbys—oak parquet or hardwood boards for the public rooms, wall-to-wall carpeting for the bedrooms and hallways, and vinyl or ceramic tile for the kitchen and bath—are no longer accepted unchallenged. Instead, there is likely to be a telling scrutiny that demands investigation of a plethora of imaginative, perhaps unlikely, alternatives.

Among wood floors alone, the variations are abundant. Parquets in a handful of designs are no longer the exclusive domain of the affluent; affordable, easy-to-install parquet tiles have reached the mass market. Now, those with greater resources can opt for one-of-a-kind custom designs that are really art underfoot. Even plain hardwoods, which for years wore some shade of golden stain, offer dramatic polarities from a deep ebony to a bleached or pickled white finish.

Soft pine floors, as well as some hardwoods, are reaping the happy results of an emerging proclivity for, and expertise in, decorative paint treatments. Although

Left: CLEVER TROMPE L'OEIL PAINTING CREATES A STONE FLOOR THAT MAKES A WITTY STATEMENT AND LINKS THE FLOOR AND THE (REAL) STONE WALL THEMATICALLY. THE CURVACEOUS FURNISHINGS AND FLORAL ACCENTS SERVE TO WARM UP THE ROOM.

wood, with its striated grain and rich patina, can certainly stand on its own, as a flooring material it can also be enlivened with paint to please a softer or more graphic taste.

Older wood floors that have withstood or succumbed to decades of aging can be painted in solid washes of color for a fresh look that conceals any scars or blemishes. Delicate stenciled borders in geometric or floral patterns, either new or historic, add an edge of interest that pulls a room together with color and pattern. Checkerboard stencils that cover an entire floor surface, and perhaps even form the pattern with squares of the natural exposed wood grain, make no attempt at subtlety but revel in the boldness of their own design statement. For a flamboyantly decorative look especially appropriate for a romantic English country cottage, an entire wood floor can be emblazoned with a floral stencil or even a freehand painting with inimitable, inventive design. Not to be forgotten—as, indeed, they represent one of the strongest trends in decorating—are decorative paint treatments such as sponging, combing, ragging, splattering, and colorwashing. Used as border trim or to decorate an entire floor, used alone or in combination with another paint treatment, each of these techniques lends a unique, handcrafted look to the floor that is a prized commodity today; the splash of color the paint treatment injects into the space is a bonus.

As if the options in wood flooring treatments aren't dizzying enough, there is still another decision to be made with this most basic flooring material: whether or not to embellish the floor further with an area rug, and if so, what kind. Oriental, rag, hooked, needlepoint, kilim, sisal, Native American woven, sculpted carpet, and even painted canvas rugs all bring exciting looks of their own to the floor. Any area rug will soften a space, cozying it up and removing a sense of hard edges. The design question is, does the room need softening? Does the area rug enhance the desired effect or does it impair it, cluttering or confusing an otherwise clean statement?

As an awareness of the value of texture and natural materials spreads from the interior design profession to the public, harder-surfaced flooring treatments are growing in popularity. As with wood floors, the possibilities for creativity here aren't stymied by a paucity of selections or limited applications. Slate, granite, marble, Pennsylvania fieldstone, Texas limestone, and other quarried stones are joining brick pavers, Mexico's stunning Saltillo tiles, darker Italian tiles, and ceramic tiles of all colors and painted decorations as serious flooring contenders. The hard-surfaced floor treatments aren't limited to entry halls or bathrooms, as in the past. Instead, they are appearing as stellar design tools in family and living rooms, studies, kitchens, libraries, solariums—virtually all of the public spaces of the home. Only expense remains a hurdle, but even that can be circumvented. Salvaged sheets of roofing slate, for example, can supply the dark charm of cut slate on the floor for a fraction of the cost, and with the right cutting and adhesive tools, even an amateur do-it-yourselfer can master the installation process.

Carpeting, like other floor treatments, requires some rethinking today. The notion of "wall-to-wall" isn't what it once was: These days, what spans the perimeters of the room, wall to wall, may be a border carpeting of one color and profile, while the main walking surface of the room is covered in an entirely different carpeting that is inset within the border trim.

Raised consumer consciousness means that the performance of carpeting is up overall, with stain-resistant, wear-resistant fibers becoming standard selling features. Natural fibers are in demand, as the prominence of 100-percent wool Berber carpet indicates. Because of its clean look and low-maintenance reality, low pile carpet remains a favorite choice not only as an industrial option but within the home, especially for contemporary spaces in which a no-fuss effect is the goal.

One of the most dramatically improved flooring materials, in an aesthetic sense, is vinyl. Vinyl sheets or components offer surprisingly high-style looks—great solid colors or even sponged looks that acknowledge the current trends in design. A handcrafted, custom look can be achieved with vinyl by creating a one-of-a-kind design, then cutting and laying components to fit the mosaic pattern.

When making any decision on flooring, some practical concerns must be addressed: the cost and availability of the materials, as well as the labor to install them (are any special skills required?); age of the home; function of the room; longevity of the flooring material; and your life-style. Decorating issues are more fun: How does the flooring work to achieve the desired mood; does it augment or diminish a sense of spaciousness; how does it merge with existing furniture and decorative style; and finally and most importantly, does it convey individuality, expressing, in some way, who you are?

Wood

With its mellow coloration and rich grain, wood brings a profound warmth and depth of character to the floor. In addition to its visual charm, wood is also one of the most practical of flooring solutions, inherently resilient in the face of scratches and scuffs and impervious to everyday wear. Not only has it retained its place as a classic flooring choice through the centuries, wood has expanded its options with a spectrum of finishes and flights of fancy in decorative patterns and paint treatments. Stenciling, faux finishes, and special paint techniques such as sponging, combing, ragging, and colorwashing lend a handcrafted, one-of-a-kind quality to wood that has only enhanced and broadened its appeal as a flooring solution. Stains, too, have become inventive. In addition to the standard golden-stained hardwood floor that dominated homes for decades, polar opposites are alluring more takers. Pale pickled or whitewashed floors create an airy, beach-house look that is especially good for casual, breezy design; rich mahogany stains anchor a home with a sense of stability and tradition; ebony creates an unmatched drama like nighttime itself—and alternated with whitewashed or natural stained boards, the effect is that of graphic art underfoot. The sheen of a wood floor is a totally controlled variable, too. For formal design, floors can be as high-gloss as water; a low-luster matte finish can allow the floor to disappear into the most subtle of white-on-white contemporary designs.

Left: In a formal entry hall, wood parquet flooring immediately establishes a rich, warm ambience in the home. Oversized squares, each bordered with a darker wood frame for a grid effect, have been stained different shades, then laid on opposing grains to produce a striking graphic pattern. **Above:** Decorative paint treatments, which are among the most popular of today's emerging decorating tools, can produce exciting effects when applied to a plain wood floor. Here, the decorative paint treatment is combing, a process in which a comb or comblike object is dragged across a still-wet painted surface to produce wavelike striations. Combing is especially effective in enhancing the slightly offbeat country character of this sewing room by lending previously unadorned wood an element of lighthearted style.

Above: MANY A SIN IS FORGIVEN ON AN OLD WOOD FLOOR WHEN THE FLOOR BECOMES A CANVAS FOR DECORATIVE STENCILING EXECUTED IN COLORFUL PAINTS. IN THIS CHILD'S BEDROOM, THE WOOD FLOOR BECAME CHEERY WITH BLOCKS OF PINK PAINT, OVER WHICH AN ELABORATE NETWORK OF STENCIL DESIGNS WAS PAINTED.

Below: CHECKERBOARD FLOOR PATTERNS ARE CLASSICS, AND WITH THE INCREASING USE OF STENCILING AS PART OF HOME DESIGN, THEY ARE APPEARING IN MORE AND MORE INTERIORS. THIS PARQUET FLOOR, LEFT NATURAL AND UNPAINTED, WOULD HAVE REMAINED YET ANOTHER NEUTRAL SURFACE IN A SPACE DEVOID OF PATTERN. STENCILED IN DRAMATIC RED AND WHITE, HOWEVER, THE WOOD FLOOR BECOMES THE SOLE SOURCE OF PATTERN AND, THUS, THE FOCAL POINT IN THE NARROW LIBRARY CORRIDOR. **Right:** WHEN THE PHYSICAL WARMTH OF WOOD AND THE COOL LOOK OF STONE ARE DESIRED, A SOLUTION LIES IN TROMPE L'OEIL PAINTWORK. THIS LIVING ROOM FLOOR IS MADE OF WOOD THAT HAS BEEN CLEVERLY DISGUISED AS FIELDSTONE WITH SOME SLEIGHT-OF-HAND BRUSHSTROKES AND PAINT. THE MESSAGE IS CLEAR: YOU CAN'T ALWAYS BELIEVE WHAT YOU SEE. THE FUN LIES IN LOOKING A LITTLE CLOSER.

Left: EVEN A FUNCTIONAL BUTLER'S PANTRY ASSUMES A PLAYFUL ATTITUDE WHEN ITS WOOD FLOORING IS INVIGORATED WITH FOOL-THE-EYE PAINTWORK. THE STONE FAUX FINISH PAINTED ONTO THE WOOD RAISES THIS QUESTION: ARE THE INDENTATIONS IN THE FLOOR DELIBERATE, PAINTED-ON BLEMISHES DESIGNED TO CONVINCINGLY APPEAR AS NATURAL IMPERFECTIONS IN THE "STONE," OR ARE THEY OLD SCARS ON THE WOOD ITSELF FROM YEARS OF SERVICE? THE QUESTION REALLY IS MOOT. THE POINT, INSTEAD, IS THAT WITH THIS PAINT TREATMENT ON WOOD, THEY WORK. **Below:** DECORATIVE STENCILING ON WOOD DATES TO COLONIAL TIMES. FOR HISTORIC HOMES OR SPACES SEEKING A YESTERYEAR, COUNTRY MOOD, THE TIME-TESTED PAINT TREATMENT ON WOOD FLOORS CREATES AN UNMISTAKABLE ERSTWHILE FEEL. IN THIS LIVING ROOM, A HISTORIC YELLOW OCHER, WHICH RESEMBLES AN EARLY, NATURAL PIGMENT PAINT WITH ITS SUBTLE, EARTHY HUE, WAS CHOSEN AS THE OVERALL FLOOR COLOR. OVER THE BACKGROUND PAINT, A SMALL, BRICK-RED STENCIL DESIGN WAS PAINTED TO ADD JUST ENOUGH DECORATION TO CREATE A HOMEY FEEL AND VISUALLY TIE THE ROOM'S COLOR SCHEMES TOGETHER.

Left: INTERIOR DESIGN IS, AT ITS BEST, IS A STUDY IN CONTRASTS. WISELY, THIS HIGHLY DECORATED BEDROOM'S WIDE-BOARD PINE FLOORS WERE LEFT IN THEIR NATURAL FORM, A WARM AND HONEST COUNTERPOINT TO YARDS OF MATCHED FABRICS AND REFINED FURNISHINGS.

Right: WHEN DECORATING TASTES RUN TO THE ECLECTIC AND FUNKY, WITH A PENCHANT FOR FOLK ART AND OFFBEAT ACCESSORIES, A WOOD FLOOR IS YET ANOTHER OPPORTUNITY TO EXTEND THE STATEMENT. THERE'S NO MISTAKING A SENSE OF PLACE WHEN "HOME" IS ANNOUNCED IN SIGNAGE, AS ART PAINTED RIGHT ONTO THE FLOOR. **Below:** A DOUBLE-TAKE IS IN ORDER IN THIS LIVING ROOM, WHERE WHAT APPEARS TO BE A HANDSOME, FINELY CRAFTED NEEDLEPOINT RUG IS ACTUALLY PAINT OVER WOOD. IT IS THIS ELEMENT OF THE UNEXPECTED THAT TRANSFORMS THE WELL-DECORATED ROOM INTO A SPACE CHARGED WITH EXCITEMENT, WHILE THE AUDACITY OF THE RICH, RED FLOOR PAINT MAKES THE ROOM REVERBERATE WITH A SENSE OF POWER.

Above: WHEN RESTRAINT IS EMPLOYED WITH FURNISHINGS AND ART, AS IN THIS ENTRY FOYER, A PALE OAK FLOOR STENCILED WITH A REPEATING VINE PATTERN TO MIMIC A NEEDLEPOINT RUG BECOMES THE DESIGN STATEMENT THAT SETS THE PITCH OF INTEREST FOR THE ENTIRE HOME.

Below: WITH THE SUN AT ITS CENTER, A CUSTOM-PAINTED WOODEN FLOOR RADIATES OUTWARD TO THE EDGES OF A HALLWAY, MAKING THIS OFTEN-IGNORED SPACE A MASTERFUL EXAMPLE OF ATTENTIVE DESIGN.

Left: WITH ITS LUSH CROWN MOLDINGS, PEDIMENT, AND CLASSICAL FIRE-PLACE, THIS FORMAL LIVING ROOM IS A NATURAL CANDIDATE FOR PLAIN PARQUET FLOORING. BUT INSTEAD OF THE EXPECTED FLOORING SOLUTION, THE OAK FLOORBOARDS HAVE BEEN STENCILED WITH A DARKER STAIN TO EMULATE A CLASSICAL PATTERN OF INLAID, PARQUET WOOD TILES, EFFECTIVELY REMOVING ANY HINT OF STALENESS FROM THE TRADITIONAL DECOR. **Above:** A SLEEK, CONTEMPORARY KITCHEN WITHOUT A TRACE OF COLOR MAINTAINS ITS NEUTRALITY WITH COMMANDING PRESENCE AS A RESULT OF ALTERNATING LIGHT AND DARK STAINS ON THE WOOD FLOOR. THE STRONG GRAPHICS CREATED BY THIS LINEAR PATTERNING GAIN EXTRA INTEREST DUE TO THE PATTERN SHIFTING DIRECTIONS MID-FLOOR AND AROUND CORNERS.

Right: THIS RICH FLOORING TREATMENT WAS CREATED BY LAYING DOWN A CHECKERBOARD STENCIL PATTERN ON A HARDWOOD FLOOR, THEN COLORING ALTERNATE SQUARES WITH EBONY AND MOTTLED BROWN STAINS. FOR THE FINAL TOUCH, A CLASSICAL GREEK KEY MOTIF WAS PAINTED AS A BORDER TREATMENT AROUND THE ROOM'S PERIMETER.

Right: ONLY A FEW DECADES AGO, HARDWOOD FLOORS INSTANTLY CONNOTED DARKNESS AND WARMTH DUE TO STANDARD GOLDEN TO DARKER BROWN STAINS THAT WERE AU COURANT. TODAY, THE HARDWOOD FLOOR CAN SUGGEST THE DIAMETRIC OPPOSITE: A LOOK AS PALE AS SAND, AS COOL AS SHALE. LIGHT STAINS, PICKLING, BLEACHING, AND WHITEWASHING ARE OPTIONS FOR GIVING HARDWOOD THE DESIRED LIGHT TOUCH.

Left: LIGHTLY APPLIED, LEAVING AN AMOUNT OF "AIR" AS EXPOSED WOOD, A BLUE STAIN INJECTS A RUSTIC, COUNTRY QUALITY TO THIS SPACE OF CONTEMPORARY ARCHITECTURE AND SPARE FURNISHINGS. IT IS THE CONTRAST OFFERED BY THE FLOORING TREATMENT THAT GIVES THE SPACE ITS CHARM. **Below:** TROMPE L'OEIL TAKES A FOLKSY TURN UNDERFOOT WITH THIS DEPICTION OF THE FAMILY DOG AT REST, PAINTED RIGHT ONTO THE HOME'S OAK FLOOR.

Above: TO ADD A DASH OF EXCITEMENT TO THIS TRADITIONAL COUNTRY DECOR, WOODEN TILES IN CONTRASTING SHADES OF LIGHT AND DARK HAVE BEEN LAID IN GROUPS OF FOUR. THE OPPOSING GRAIN ARRANGEMENT PROVIDES FURTHER INTEREST.

Above: MUCH OF THE EXCITEMENT IN INTERIOR DESIGN TODAY EMANATES FROM RETHINKING DECORATING RULES—FOR EXAMPLE, USING WOOD AS A VIABLE FLOORING MATERIAL FOR A SLEEK, SOPHISTICATED CONTEMPORARY BATHROOM. LAID ON THE DIAGONAL (A TREATMENT THAT CARRIES MORE VISUAL INTEREST THAN SQUARING OFF AT THE ROOM PERIMETERS), WOOD TILES IN TWO SUBTLE STAINS CREATE A CHECKERBOARD PATTERN THAT COMPLETES THE UNDERSTATED BUT ELEGANT THEME OF THE SPACE. **Right:** THE OLD HERRINGBONE DESIGN THAT WAS ONCE CONSIDERED SYNONYMOUS WITH PARQUET FLOORS HAS SOME COMPETITION THESE DAYS, AS EVIDENCED BY THIS INTRICATE V-SHAPED MOTIF THAT IS A SURPRISINGLY COMPATIBLE PARTNER WITH THE HOME'S TURN-OF-THE-CENTURY CRAFTSMAN ARCHITECTURE.

HARDY NATURALS

As consciousness of the earth and the conservation of its precious resources rises, so does the desire to live with natural materials that bring a little of the outdoors inside the home. Stone floors or tile floors made from baked earth are celebrated features in many homes, and today they are found in more rooms within those homes. For cool good looks, nothing is more effective than a slate, granite, marble, Pennsylvania fieldstone, or Texas limestone floor. Handsome terra-cotta floors of Saltillo tiles inject a regional southwestern flavor to a home, while brick pavers, with their oxblood color and rough texture, set the imagination traveling down quainter, vintage avenues. Veined marble as a floor treatment announces sophisticated living; creamy travertine can suggest neoclassical or natural contemporary style; and irregular shapes of fieldstone on the floor set the stage for an Adirondacks-style lodge look. In other words, a

hard-surfaced natural material can be found to answer almost any decorating need. The only limitations are expense and other practical considerations such as cold and noise factors. For visual and textural interest, the naturals remain unchallenged.

Left: Textured, natural stone tiles bring a welcome rusticity to an otherwise streamlined bathroom; their unexpected warmth prevents the room from appearing sterile. **Above:** Finely polished marble, a time-tested flooring option, is a foolproof design tool for imbuing the bathroom with opulent presence. Here, the marble lends easy-on-the-eye unity to the space by extending from the floor, in stairstep ascension, to encase a luxurious whirlpool bath.

Below: WHITE CERAMIC TILES LAID IN CONFIGURATIONS OF FOUR, THEN BORDERED WITH SMALLER, TEAL TILES WITH A SINGLE LIGHTER ACCENT, PROVIDE ALL THE NECESSARY PATTERN IN THIS BATHROOM. THE TILE THEME IS REPEATED AT THE CHAIRRAIL AND INSIDE THE SHOWER TO CREATE A HARMONIOUS SPACE.

Above: BAKED ITALIAN TILES BRING THE AMBIENCE OF A TUSCAN VILLA TO THE INTERIOR OF AN URBAN APARTMENT. **Right:** TERRA-COTTA TILES, MORE THAN MERE FLOORING, MAKE A POWERFUL DESIGN STATEMENT IN THIS BATH-ROOM THAT HAS A STARBURST PATTERN AS ITS CENTERPIECE. THE TILES PROVIDE A CHARMING FOIL FOR THE ROOM'S NEOCLASSICAL TOUCHES, SUCH AS THE UNDULATING, POOLED FABRIC CURTAINS USED TO DEFINE THE SINK AREA.

Left: REMNANT CERAMIC TILES DON'T HAVE TO BE TOSSED IN THE TRASH. THEY CAN FIND NEW LIFE WHEN ASSEMBLED TOGETHER IN A CRAZY-QUILT FLOORING PATTERN OF SORTS — A FUNKY EFFECT THAT'S RIGHT FOR THIS SIMPLE WHITE BATHROOM WITH ITS FIRE-ENGINE RED SINK.

Left: USUALLY ASSOCIATED WITH AMERICAN SOUTHWESTERN, MEXICAN, OR SOUTHERN EUROPEAN STYLE, TERRA-COTTA TILES OF A DARK SIENNA HUE PROVIDE NOTHING SHORT OF DRAMA WHEN TEAMED WITH THE SHEER WHITE PLANES AND GLASS EXPANSES OF CONTEMPORARY DESIGN, CALLING ATTENTION TO THE ARCHITECTURE WHILE CUTTING ITS POTENTIALLY COLD EDGE.

Right: WITH ITS MEMPHIS COLORS OF SCREAMING YELLOW AND BLACK, THIS KITCHEN GAINS STABILITY AT THE ROOTS WITH A CERAMIC TILE FLOOR GLAZED A UNIFORM NEUTRAL.

Left: LONG ACCEPTED AS A SENSIBLE FLOORING SOLUTION FOR WET SPACES SUCH AS THE KITCHEN AND BATH, CERAMIC TILES EXPRESS A DIFFERENT SORT OF SENSIBILITY IN THIS CONTEMPORARY KITCHEN, WHERE THEY NOT ONLY ADORN THE FLOOR BUT CLIMB THE WALL TO THE CEILING FOR A CHIC LOOK. **Right:** FIELDSTONE, A BUILDING MATERIAL ASSOCIATED WITH RUSTIC ROCKY MOUNTAIN LODGES AND ADIRONDACKS GREAT CAMPS, CAN GIVE EVEN A DETERMINEDLY MODERN SPACE AN OUTDOORSY AIR.

Left: Painted wood isn't the only means of achieving trickery underfoot. In this lavish Mediterranean-style corridor, terra-cotta tiles are disrupted by what appears to be a richly hued, finely woven Oriental rug. In fact, the rug is also made of tiles, custom designed to create a convincing rug design when properly laid on the floor.

Right: Painted ceramic tiles can lend a romantic, cottage-country ambiance to a space, as proven by this bathroom's floor treatment. Only the briefest of valances at the window is necessary to underscore the vitality of the cheery floor.

Left: A Byzantine flavor is imparted by tiny ceramic tiles intricately laid into the dark marble floor to form a mythological mosaic—a design replete with grandeur and classic style.

Right: GRAY GLAZED CERAMIC TILES WITH A LIGHTER CAULKING SERVE AS A TRANSITIONAL DEVICE IN THIS KITCHEN, WHICH COMBINES SOUTHWESTERN MOTIFS, CONTEMPORARY FURNISHINGS, AND AN ART DECO-STYLE, BLACK-AND-WHITE TILE BACKSPLASH. **Below, right:** EVEN IF THE COMPLEXITIES OF LIFE CAN'T BE REDUCED TO BLACK AND WHITE, THE ENVIRONMENT IN WHICH WE GRAPPLE WITH THEM CAN BE. THIS KITCHEN FLOOR UTILIZES BLACK-AND-WHITE CERAMIC TILES, INTERMITTENTLY DISPERSED, TO CREATE A SPACE THAT IS CLEAN AND GRAPHICALLY APPEALING, WITH ECHOES OF AN ART DECO THEME.

Right: FORGET THE IDEA THAT BRICK PAVERS ARE LIMITED TO STREETS. USED AS FLOORING MATERIAL FOR THE HOME, THEY CONVEY WARMTH, TEXTURE, AND RUDDY BEAUTY. IN THIS ELEGANT YET RUSTIC ENTRY, PAVERS OF TWO DIFFERENT COLORS HAVE BEEN PLACED AT IRREGULAR DIAGONALS TO HEIGHTEN THE INTEREST.

Below: CERAMIC TILES AREN'T JUST FOR THE KITCHEN OR BATH. THEY CAN PROVIDE A VIBRANT, VISUALLY INTRIGUING FLOORING SOLUTION FOR MAJOR LIVING AREAS OF THE HOME, AS WELL. IN THIS LOFT FAMILY ROOM, BLACK AND WHITE TILES LAID ON THE DIAGONAL IN A CHECKERBOARD PATTERN PRESENT THE ROOM'S PATTERN AND RHYTHM, ALLOWING OTHER FURNISHINGS TO ASSUME A MORE LOW-KEY ROLE SUITED TO THE DESIGN'S CONTEMPORARY GENRE.

Above: MEXICAN SALTILLO TILES, WITH THEIR TERRA-COTTA COLOR AND RIPPLE-TEXTURED SURFACE, CREATE INTENSE FLOOR INTEREST. NOT ONLY DO THEY IMPART A HISPANIC SENSIBILITY TO THIS LIGHT, AIRY SPACE, THEY ALSO PROVIDE A NATURAL FEELING UNDERFOOT, GIVING ONE A SENSE OF WALKING UPON THE EARTH ITSELF. **Left:** DRAWING FROM THE CONCEPT OF PATTERNED FABRICS, THIS CERAMIC-TILE KITCHEN FLOOR UTILIZES A GROUND COLOR (BLACK) AS THE DOMINANT HUE, THEN ADDS EXCITEMENT WITH DESIGNS ATOP THAT NEUTRAL FIELD. IN THIS CASE, THE MOTIFS ARE CREATED WITH SMALLER-CUT WHITE AND BLACK LEAF MOTIF TILES. SMALLER-CUT WHITE AND BLACK LEAF-MOTIF TILES. THE CURVED SHAPES ECHO THE ARCHITECTURE OF THIS DRAMATIC ROOM.

TEXTILES

Due to its physical warmth, visual softness, noise reduction, and tactile comfort, carpet retains a unique place among floor treatments. On a cold winter morning, no flooring material is more welcoming to bare feet than carpet's warm, cushioning touch. Another practical advantage is the seamless, streamlined look a wall-to-wall carpeted floor can lend to a space. Carpet also has a low-maintenance quality, camouflaging dust more thoroughly than a hard-surfaced flooring material. Improved stain-resistant treatments for fibers have made carpeting increasingly impervious to spills and stains, though it still is no rival for the easy wipe-up of a hard surface.

Subject to style and color trends, carpet can quickly date a house, as rooms still bearing vestiges of fiery orange shag attest. But, at the same time, new carpet can instantly imbue a home with up-to-the-minute panache. The most appealing selections currently include natural wools, loopy Berbers, low-pile plushes and industrials, and custom creations that include a border trim or design of a different color or type than the main carpeting.

A compromise between hard-surfaced and wall-to-wall carpeted floors, the area rug is a versatile alternative with something to offer in every decorating style. Offerings range from formal, traditional Orientals and Aubussons to primitive rag, hooked, and painted canvas rugs, with contemporary sisals and kilims and a host of other handsome types in between—an option is available to suit virtually every personal taste.

Above: Permitting the best of both worlds, a wool staircase runner and matching area rug provide the physical and visual warmth of carpeting, while still allowing the natural beauty of a marble entry and hardwood stairs to peek out around the peripheries. **Left:** The plain weave of sisal, used wall-to-wall in this streamlined, contemporary space, provides a seamless look that underscores the room's quiet elegance. Sisal's golden hue complements the room's dark woods.

Below: WHEN A PRIMITIVE, HOMESPUN EFFECT IS DESIRED FOR A HISTORIC SPACE, A SIMPLE RAG RUG TOPPING A WOOD FLOOR IS A NATURAL. THIS RUG CONVEYS THE NOTION OF HANDCRAFTSMANSHIP AND SIMPLICITY, WHILE GENTLY INTRODUCING THE IDEA OF COLOR. THE TEXTILE ALSO SERVES TO SOFTEN THE HARDNESS OF THE ANTIQUE WOOD FURNISHINGS.

Above: FOR A CHILD'S BEDROOM OR PLAYROOM, AN AREA RUG IS OFTEN THE BEST SOLUTION WHEN THE GOAL IS TO REPEAT A PRESELECTED OVERALL ROOM THEME UPON THE FLOOR. HERE, THE MENAGERIE REPRESENTED IN THE MURAL MARCHES ACROSS THE FLOOR WITH GREAT STYLE IN A SUBTLY PATTERNED AREA RUG CHIC ENOUGH FOR ADULTS TO ENJOY, TOO. **Right:** NATURAL FIBERS ARE VERSATILE ENOUGH TO ENHANCE A WIDE RANGE OF DECORATING STYLES, FROM CASUAL TO FORMAL. THE SIMPLE GEOMETRIC MOTIFS ALONG THE BORDER OF THIS FINELY WOVEN RUG ECHO THE SMALLER PATTERN OF THE CURTAINS, CREATING HARMONY AMONG THE DIFFERENT ELEMENTS OF THE ROOM.

Left: THE OLD NOTION OF CARPETING EXTENDING WALL-TO-WALL IS RECONSIDERED IN THIS GREAT ROOM, IN WHICH CUSTOM CARPETING DEFINES A RAISED-PLATFORM SITTING AREA, DIFFERENTIATING IT FROM THE OAK-FLOORED DINING AREA. **Below:** NEEDLEPOINT AREA RUGS WERE A STAPLE IN EARLY HOMES FOR OBVIOUS REASONS: THEY IMBUE AN INTERIOR WITH DELICATE CHARM, OLD-WORLD ROMANCE, AND FINELY WORKED BEAUTY. THEIR GENTLE MOTIFS AND SOFT, PASTEL COLORS ARE IDEAL FOR A BEDROOM, WHERE SERENITY IS THE FIRST ORDER OF BUSINESS.

Left: AMONG THE MOST POPULAR MOTIFS IN HAND-NEEDLE-POINTED RUGS ARE FLORAL DESIGNS. IN THIS ROMANTIC VICTORIAN BEDROOM, CABBAGE ROSES BLOOMING ON THE AREA RUG UNDERSCORE THE SOFTNESS OF DUST-RUFFLES, SHAMS, AND A MINI-PRINT WALLCOVERING.

Right: AREA RUGS CAN
PROVIDE INFORMAL DOLLOPS OF
COLOR AND TEXTURE THAT SERVE AS
ACCENTS IN A ROOM, OR THEY
CAN SERVE AS FORMAL PLATFORMS
THAT GROUND AND DEFINE THE
SPACE ITSELF, AS IN THIS RUG. THE
COLORS IN THE RUG PROVIDE THE
PALETTE FOR THE ROOM SCHEME,
WHILE THE RUG DESIGN, WITH ITS
SPARSE BACKGROUND MOTIF AND
MORE ELABORATE BORDER DECORA-
TION, ESTABLISHES THE ROOM'S
DESIGN PHILOSOPHY OF UNDER-
STATED ELEGANCE.

Right: DESIGNED IN A MULTIFARIOUS PALETTE OF RICH BURGUNDY, INDIGO,
AND PLUM, THIS RAG RUG WARMS THE RUSTIC BEDROOM WITH ITS ENVELOPING
BLANKET OF COLOR. LOOMED TO AN OVERSIZED PROPORTION, IT SPANS THE
BEDROOM ALMOST TO THE WALLS, CREATING THE COCOONING EFFECT OF
WALL-TO-WALL CARPETING, BUT WITH THE TEXTURAL INTEREST AND VARIEGATED
COLOR THAT ONLY A RAG RUG CAN PROVIDE.

Left: WOOL CARPET DOESN'T HAVE TO STRETCH FROM WALL TO WALL, NOR IS THERE ANY RULE THAT SAYS IT MUST BE RECTILINEAR IN FORM. IN THIS DINING ROOM, THE CARPET IS CUT TO A SERPENTINE SHAPE THAT ADDS FLUIDITY TO THE SHARP GEOMETRIC EDGES OF THE DINING TABLE AND CHAIRS. THE TENSION BETWEEN THE RUG'S CURVACEOUS FORM AND THE STRAIGHT LINES OF THE FURNISHINGS PROVIDES THE SOURCE OF VISUAL INTEREST FOR THE SPACE. **Above, left:** THE IDEA OF WALL-TO-WALL CARPETING AS A SEAMLESS SEA OF SOFT TEXTILES UNDERFOOT IS VALID, BUT NOT ETCHED IN STONE. WHEN A DIFFERENT EFFECT IS DESIRED— ONE THAT BREAKS UP THE MONOTONY OF THE LONG STRETCH OF UNINTERRUPTED, SAME-PILE CARPETING—THE RESULTS CAN TAKE A CREATIVE TURN THAT PUTS A NEW SPIN ON THE BASIC PREMISE OF THE FLOORING MATERIAL. IN THIS FAMILY ROOM, CARPETING EXTENDS TO THE WALLS, BUT IS CUT INTO SLICES THAT RESEMBLE JAGGED BLOCKS OF STONE. **Above, right:** FOR SHEER CREATURE COMFORT, NOTHING CAN COMPETE WITH PLUSH WALL-TO-WALL CARPETING THAT BECKONS PEOPLE TO KICK OFF THEIR SHOES AND INDULGE IN ONE OF THE SIMPLE PLEASURES OF DAILY LIVING. THIS CARPET'S NEUTRAL TONE ENHANCES THE SOOTHING, MODERN LINES OF THIS BEDROOM.

Below: THE CLASSIC MOTIF OF STRIPES CAN TAKE A NUMBER OF FORMS, FROM UPSCALE AND DRESSY TO DOWN-HOME AND CASUAL. IN THIS SUNNY ATRIUM, A WOVEN, NATURAL-FIBER RUG FEATURES A WIDE STRIPE DESIGN, ALMOST MIMICKING THAT OF AN AWNING, THAT SERVES TO ANCHOR THE SPACE IN CASUAL, INVITING STYLE. **Right:** SISAL MATTING, WITH ITS NATURAL FIBERS, WARM COLOR, AND RUSTIC SIMPLICITY, HAS BECOME AN EXTREMELY POPULAR CHOICE FOR COUNTRY AND CITY HOMES. IN THIS COUNTRY SPACE, THE EMPHASIS IS ON SISAL'S PRACTICALITY. STRETCHED ALL THE WAY TO THE ROOM'S EDGES, IT PROVIDES VISUAL CONTINUITY AND EASY MAINTENANCE.

Right: SISAL, WITH ITS CRISP YET WARM WEAVE, HAS BECOME A MODERN FAVORITE IN HOME DESIGN NOW THAT TEXTURE HAS JOINED COLOR AND PATTERN AS A MAJOR DECORATING OBJECTIVE. A VERSATILE FIBER, IT CAN BE WOVEN INTO A FINE, TIGHT PATTERN OR A LOOSE LOOPY WEAVE. IN THIS LIVING ROOM, THE GOAL WAS TO CREATE A SPACE THAT BECKONED WITH COMPELLING, STAY-AWHILE COMFORT. THE NUBBLY TEXTURE OF THE WEAVE ENHANCES THAT EFFECT FROM THE BOTTOM UP.

Below: ORIENTAL CARPETS, WITH THEIR EXQUISITE CRAFTSMANSHIP, BOUNTIFUL COLOR, AND ELABORATE DESIGN, HAVE BEEN A MAINSTAY OF FINE HOMES THROUGHOUT THE WORLD FOR CENTURIES. THEIR USE IS NOT LIMITED TO TRADITIONAL DECOR, HOWEVER. THE FINENESS OF AN ORIENTAL AREA RUG CAN IMBUE A CONTEMPORARY SPACE IN WHICH THE DOMINANT COLORS ARE COOL WITH A DEPTH OF CHARACTER AND WARMTH, AS ILLUSTRATED BY THE RICHLY PATTERNED ORIENTAL IN THIS LIVING ROOM.

Left: IN THE HANDS OF A CREATIVE DESIGNER, FIBER KNOWS NO BOUNDS AS A FLOORING TREATMENT. IN THIS COLOR-PACKED CONTEMPORARY ROOM, A GLOBAL CONSCIOUSNESS IS PLAYFULLY SUGGESTED BY AN AREA CARPET CUT IN THE SHAPE OF A CONTINENT. THE DRAMATIC SWIRLING PATTERN OF COOL GREENS, PURPLES, AND BLUES TAKES THE COLOR WHEEL ON FULL SPIN, COUNTERING THE HOT REDS AND GOLDS OF THE FURNITURE. **Above:** A CONTEMPORARY ROOM'S SLEEK BACKGROUND FLOORING TREATMENT WAS FIRST ESTABLISHED WITH A COOL, GRAY EXPANSE OF CONCRETE. OVER THAT, A WEDGE OF CONTEMPORARY DESIGN—AN AREA RUG REMINISCENT OF ONE OF MATISSE'S FAMOUS CUTOUTS—SERVES AS AN *OBJET D'ART* ON THE FLOOR. THE RUG'S VIBRANT BLUE AND SPIRAL SHAPE SOFTENS THE HARSHNESS OF THE GRAY AND BLACK COLOR SCHEME.

Vinyls

An increased awareness of design isn't limited to home decorating itself. The design ethic has permeated all consumer products for the home, with manufacturers getting smarter about the importance of products that look good and function well in addition to coming in at the right price point. Prominent beneficiaries of this emerging design sensibility are vinyl flooring products. Only the snootiest or most uninformed decorator continues to malign vinyl as a less-than-stylish flooring solution. Instead of a handful of stamped patterns that announce low-end stature, vinyl is now offered in designer colors, patterns that emulate faux-finish paint, and components that can be

customized as cheery, dramatic checkerboard patterns or more sophisticated one-of-a-kind designs created in mosaic fashion. To its new good looks, vinyl adds its continuing value, durability, and low maintenance, making it a flooring choice no longer relegated to the kitchen, mud room, or bath, but appearing more and more often in the main living areas of the home.

Left: In step with consumers' growing awareness of design issues and their increasingly sophisticated tastes, vinyl flooring no longer is limited to a few staple motifs and colors. Vinyl, in fact, no longer necessarily has any pattern at all, but can take the form of a sheer expanse of vivid color, as proven by this bright yellow vinyl sheet flooring. It offers the same colorful impact as a painted flooring treatment, but requires less maintenance and is lower in cost. **Above:** Like other flooring solutions today, vinyl can be inventive — as customized as the homeowner's creativity permits. For this country kitchen, the idea was to create a floor that underscored the homey Americana theme. Vinyl components featuring geometric quilt motifs were interspersed with solid-color vinyl blocks in taupe and off-white to produce a floor that bespeaks quintessential country design.

Above: THE EXCITEMENT IN ANY HOME'S INTERIOR DESIGN EMANATES FROM ITS RISK FACTOR—FROM ITS DARING TO CHALLENGE ORTHODOX METHODS AND EXPECTED SOLUTIONS. IN THIS JAPANESE-FLAVORED CONTEMPORARY HOME, THE BENT RULE LIES IN THE BOLD DECISION TO EMPLOY VINYL SHEET FLOORING, NOT IN THE EXPECTED "WET" SPACES OF THE HOME, BUT AT THE FRONT DOOR. INSTEAD OF THE ANTICIPATED PALE, PICKLED WOOD FLOOR, A TAUPE VINYL SHEET WITH A TINY V-MOTIF IN A DARKER SHADE GREETS VISITORS AT THE ENTRY AND EXTENDS INTO THE MAIN PUBLIC SPACES OF THE HOME. THE LOOK IS CONTEMPORARY, CLEAN, AND QUIETLY DRAMATIC. **Right:** IN THIS CONTEMPORARY SPACE, NATURE IS THE SPRINGBOARD FOR DESIGN SOLUTIONS. THE FLOOR HAS THE LOOK OF MARBLE, BUT TO ACHIEVE DURABILITY AND COST SAVINGS, THE DESIGNERS ACTUALLY USED MARBLE-LOOK VINYL. THERE COULD BE NO BETTER PROOF THAT TODAY'S CHOICES IN VINYL PATTERNS HAVE BEEN GREATLY EXPANDED TO INCLUDE PRODUCTS PLEASING TO EVEN THE MOST SOPHISTICATED SENSIBILITIES.

Left: THE MATTE LOOK OF NATURAL TRAVERTINE IS ACHIEVED FOR A FRACTION OF THE COST ON THIS KITCHEN FLOOR WITH VINYL COMPONENTS. WITH ITS SUBTLE COLORATION, THE FLOORING ALLOWS THE WARMTH OF THE NATURAL BUTCHER BLOCK COUNTER AND ISLAND TO TAKE CENTER STAGE.

Left: THE CLASSIC VINYL FLOORING PATTERN OF RED BRICK MOSAIC THAT WAS POPULAR DECADES AGO STILL BELONGS IN HOMES TODAY. BUT INSTEAD OF APPEARING IN EVERY NEWLY CONSTRUCTED TRACT HOME'S KITCHEN, THE VINYL BRICK FLOORING FINDS ITS MORE NATURAL NICHE IN VINTAGE, PRIMITIVE HOMES SUCH AS THIS ONE, IN WHICH EXPOSED CEILING BEAMS PRODUCE EARLY, RUSTIC CHARM AND PERIOD COUNTRY FURNISHINGS COMPLETE THE HEARTY AMBIENCE.

Right: VINYL SHEET FLOORING WITH A HIGH-GLOSS FINISH SO SHINY IT LOOKS ALMOST LIKE MICA IS A COOL COUNTERPOINT TO THE EARTHY CHARACTER OF THIS KITCHEN WITH ITS CUSTOM WOOD CABINETRY AND TILE COUNTERTOPS. THE ABSENCE OF PATTERN IN THE FLOORING MATERIAL ALLOWS THE RICHNESS OF THE WOOD TO MAKE A STATEMENT WITHOUT COMPETITION, WHILE THE FLOORING'S SUBTLE TEXTURE LENDS SUFFICIENT INTEREST TO KEEP THIS ELEMENT OF THE ROOM FROM BECOMING BLAND.

Left: A MEMPHIS-STYLE FLOOR, WITH AN ACCENT ON BRIGHT, PRIMARY COLORS AND GEOMETRIC SHAPES, PLUS AN IRREGULARITY OF PATTERNING, WAS ACHIEVED IN THIS KITCHEN USING REMNANT SQUARES OF VINYL COMPONENT FLOOR TILES. COLORS NOT TYPICALLY COMBINED WORK WELL HERE, IN A PATTERN THAT MIXES REGULAR SQUARES WITH SMALLER PIECES FOR A FLOOR THAT PULSATES WITH A JAZZ-LIKE RHYTHM. **Right:** LITTERED WITH SAILBOATS AND SEASHELLS, THIS OCEAN-INSPIRED BEDROOM APPEARS AFLOAT ON THE DEEP BLUE, THANKS TO RICHLY SATURATED BLUE VINYL FLOORING. THE COLOR CONTRAST BETWEEN THE DEEP BLUE VINYL AND CRISP WHITE WICKER MAKES THE FURNISHINGS APPEAR SUSPENDED ON WATER.

Left: FOR A GARDEN ROOM THAT OPENS ONTO THE OUTDOORS, VINYL IS A PRACTICAL FLOORING MATERIAL. THIS SPACE IS PROOF THAT A PRACTICAL CHOICE CAN SCORE HIGH IN AESTHETICS, TOO. THE VINYL TREATMENT SHOWN HERE TAKES THE CLASSIC FLOOR MOTIF OF CHECKERBOARD, THEN SHUFFLES THE GAME A BIT, MAKING SOME CHECKS HUGE, OTHERS SMALL, AND TURNING YET OTHERS ON END FOR A QUIRKY EFFECT THAT BRINGS AN ARTISTIC TOUCH TO THE FLOOR.

Left: ONE OF THE MOST PROMISING EMERGING USES OF VINYL FLOORING ENTAILS A LITTLE GOOD-NATURED DECEIT, WITH VINYL COMPONENTS POSING AS AREA RUGS. THE BENIGN IMPOSTER IN THIS FAMILY ROOM IS A GEOMETRIC VINYL "CARPET" REPLETE WITH A COLORFUL BORDER DESIGN AND NEUTRAL CENTRAL BACKGROUND. WITH THIS CREATIVE INTERPRETATION OF VINYL, THE SHARP LOOKS OF A TEXTILE RUG ARE ACHIEVED WITH NONE OF THE HIGH MAINTENANCE.

Below: IN THIS KITCHEN, THE VISUAL GOAL WAS TO CREATE THE LOOK OF A FIELD OF WHITE CERAMIC TILES INJECTED WITH COLORFUL, ABSTRACT SQUARES. THE PRACTICAL GOAL, HOWEVER, WAS TO AVOID THE HIGH MAINTENANCE OF GLAZED CERAMIC TILES, ESPECIALLY THE CLEANING OF GROUT. BOTH GOALS WERE ACHIEVED WITH VINYL FLOORING THAT EMULATES THE LOOK OF CERAMIC TILES.

Above: THIS GARDEN BREAKFAST SPACE ACHIEVES THE LOOK OF ITALIAN TILE WITHOUT THE EXPENSE AND PHYSICAL COLDNESS. THANKS TO A CONTINUAL EFFORT TO IMPROVE DESIGN AND OFFER MORE OPTIONS IN PATTERN, VINYL CAN COMPLEMENT EVEN THE MOST SOPHISTICATED DECORS. **Right:** A LOW-LUSTER, MATTE FINISH ON THIS VINYL FLOOR ENHANCES THE SOFT UNDERSTATE-MENT OF THIS STUDY IN WHITE. THE CHECKED PATTERN ON THE FLOORING INCORPORATES STRIPES IN ITS DESIGN, ENABLING THE ROOM TO INCLUDE BOTH MATTRESS TICKING AS UPHOLSTERY ON THE WICKER CHAIR AND CHECKED PATTERNS ON ACCESSORIES.

Left: ONE OF THE MOST FLEXIBLE FLOORING MATERIALS ON THE MARKET, VINYL CAN BE CUT TO ANY SHAPE, WRAPPING AROUND STAIRCASES, OUTLINING CORNERS, DEFINING ANY AREA OF THE ROOM. IN THIS FORMAL ENTRY, VINYL'S PLIANT NATURE HAS BEEN WORKED TO OPTIMUM ADVANTAGE, DRESS-ING UP THE HOME'S FLOOR WITH CREATIVE EXPRESSIONS OF STYLE.

Left: VINYL TAKES A TURN FOR THE DRAMATIC IN THIS BATHROOM, WHERE A DUSKY, SLATE-LIKE HUE SATURATES THE VINYL AND CREATES THE IMPRESSION OF HARD TILES. MORE AND MORE, VINYL, IN ITS NEW, HIGH-STYLE VARIATIONS, IS AUGMENTING ELEGANT SPACES THAT FEATURE HIGH-END FIXTURES AND FURNISHINGS.

Below: LONG RECOGNIZED FOR ITS EASY, MOP-UP MAINTENANCE, VINYL FLOORING REMAINS A FAVORITE SOLUTION FOR KITCHEN FLOORS. THIS DIAMOND-PATTERN VINYL IN NEUTRAL COLORS CREATES JUST ENOUGH VISUAL INTEREST TO ENHANCE THE SPACE WITHOUT DETRACTING FROM THE OVERALL THEME OF UNCLUTTERED SIMPLICITY.

Above: THE ARCHITECTURAL BEAUTY OF FINELY DETAILED, WHITE-PAINTED WOOD CABINETRY IN THIS AIRY KITCHEN IS GIVEN AN APPROPRIATELY CLEAN-LINED, WELL-DESIGNED FOUNDATION, WITH A CUSTOM VINYL FLOORING TREATMENT THAT INCORPORATES A GEOMETRIC DESIGN IN NAVY AND WHITE OUTLINING THE KITCHEN'S CENTER ISLAND. **Right:** EXUBERANT STYLE THAT ESCHEWS BUSY PATTERN IS CREATED IN THIS KITCHEN WITH A VINYL SHEET FLOOR IN COBALT BLUE. THE DEEP BLUE WORKS WITH ITS COMPLEMENTARY COLOR, PRIMARY YELLOW, ON THE DINING TABLE TO CREATE A SPACE THAT'S ALIVE WITH VIBRANT COLOR AND A PLAYFULNESS SUGGESTIVE OF A CHILD'S PACK OF CRAYONS.

DESIGNER SOURCES

Floors

A B C Carpet and Home
888 Broadway
New York, NY 10003
(212) 473-3000
(carpeting)

Armstrong World Industries,
Inc.
P.O. Box 3001
Lancaster, PA 17604
(717) 397-0611

6911 Decarie Blvd.
Montreal, Quebec
H3W 3E5 Canada
(514) 343-8178
(vinyl)

Beauceville Flooring, Inc.
P.O. Box 116
Beauceville West, Quebec
G0M 1A0 Canada
(418) 774-3365
(hardwood)

Bruce Hardwood Floors
16803 Dallas Parkway
Dallas, TX 75248
(800) 722-4647
Canada (800) 334-4064
(hardwood)

Cangoleum Corporation
211 University Plaza 2
3705 Quakerbridge Road
Mercerville, NJ 08619
(800) 934-3567
(vinyl)

Country Floors
15 East 16th Street
New York, NY 10003
(212) 627-8300
(ceramic tile)

Einstein Moomjy Inc.
20 Hook Mountain Road
Pine Brook, NJ 07058
(201) 575-0895
(carpeting)

Florida Tile Industries, Inc.
P.O. Box 447
Lakeland, FL 33802
(813) 687-7171
(tile)

Hartco
Tribbals Flooring Co.
900 South Gay Street
Suite 2102
Knoxville, TN 37902
(615) 544-0767
(hardwood)

Interceramic, USA
1624 West Crosby Road
Suite 120
Carrollton, TX 75006
(800) 365-6733
(tile)

J.L. Powell & Co., Inc.
600 South Madison Street
Whiteville, NC 28472
(800) 227-2007
(antique parquet floors)

Mannington Floors
P.O. Box 30
Salem, NJ 08079
(609) 935-3000
(vinyl)

Stanley Knight Ltd.
226 Boucher Street East
Meaford, Ontario
N4L 1B7 Canada
(519) 538-3300
(hardwood)

Trojan Board Limited
333 Archibald Street
St. Boniface, Winnipeg
R2J 0W6 Canada
(204) 233-7171
(hardwood)

Designers

(page 6)
Judy Dodd
Napa, CA
(707) 257-2815

(page 10)
Rubén de Saavedra
New York, NY
(212) 759-2892

(page 11)
Beverly Ellsley Interiors
Westport, CT
(203) 227-1157

(page 12, right)
Carleton Varney
Dorothy Draper & Co., Inc.
New York, NY
(212) 758-2810

(page 14)
Manuel de Santaren
Boston, MA
(617) 367-4332

(page 16, left)
Dianne Warner
New York, NY
(212) 319-7283

(page 16, right)
Rena Fortgang, designer
Locust Valley, NY
(516) 759-7826

Frank Torres, painter
New York, NY
(212) 874-2823

(page 17)
Lee Ames
Cold Spring Harbor, NY
(516) 692-8779

Diane Kovacs
Roslyn, NY
(516) 625-0703

(pages 18 and 20)
Gary Crain
Gary Crain Associates, Inc.
New York, NY
(212) 223-2050

(page 19, left)
Michael Chaves
Michael Chaves Advertising
New York, NY
(212) 677-5480

(page 19, right)
Sam Botero
Samuel Botero Associates
New York, NY
(212) 935-5155

(page 21)
Martin Kuckly
New York, NY
(212) 772-2228

(pages 22; 31; 36, bottom; 57)
Brian Murphy
Santa Monica, CA
(310) 459-0955

(page 23, top)
Janet Lohman
Los Angeles, CA
(310) 471-3955

(page 23, bottom)
Bonnie Siracusa, artist
Great Neck, NY
(516) 482-3349

(page 24)
Michael Berman
Los Angeles, CA
(213) 655-9813

(page 26)
Ron Goldman, Architect
Malibu, CA
(310) 456-1831

(page 27)
Lisa Pontillo
Lisa Pontillo Interiors
Great Neck, NY
(516) 487-0050

(page 29)
Marilee Schempp
DESIGN 1
Summit, NJ
(908) 277-1110

(page 30)
Kurth and Kurth
Mt. Kisco, NY
(914) 666-0580

(page 32, top)
Gerald Kuhn
New York, NY
(212) 889-7599

(page 32, bottom)
Stephen Paul Ackerman
New York, NY
(212) 938-1260

(page 35, top)
Peter LaBau of Classic
    Restorations
Cambridge, MA
(617) 492-1603

(page 35, bottom)
Stephen Huberman
Stephen & Gail Huberman
Woodbury, NY
(516) 364-1770

(page 36, top)
Bogdanow & Associates,
    Architects
New York, NY
(212) 966-0313

(page 40)
Ralph Lauren
Polo/Ralph Lauren
New York, NY
(212) 318-7130

(page 42, right)
Herborg McLaughlin
Herborg Interiors
West Orange, NJ
(201) 325-3605

(page 43)
Tom Dennis Design
Chicago, IL
(916) 455-2696

(page 44)
Betty Sherrill
McMillen Inc.
New York, NY
(212) 753-5600

(page 45, top)
Demetri Sarantitis, Architect
New York, NY
(212) 925-8285

(page 45, bottom)
Peter van Hattum
Van Hattum & Simmons, Inc.
New York, NY
(212) 593-5744

(page 46)
Jeff Haines
Butler's of Far Hills
Far Hills, NJ
(908) 234-1764

(page 47)
Barbara Ostrom
Barbara Ostrom Associates
Mahwah, NJ
(201) 529-0444

(page 48)
Neil Korpinan
Los Angeles, CA
(213) 661-9861

(page 49, left)
David Livingston Interiors
San Francisco, CA
(415) 392-2465

(page 49, right)
Roy McMakin
Seattle, WA
(206) 323-6992

(page 50, both)
Anita Calero
New York, NY
(212) 727-8949

(page 52)
Steven Harris
New York, NY
(212) 587-1108

(page 53, right)
Theodore & Theodore,
    Architects
Dresden, ME
(207) 737-2131

(page 54)
Ann Fox
Dallas, TX
(214) 369-7666

(pages 55; 58; 60, top; 61;
    62; 63; 64, both; 65;
    66; 67; 68, top)
Armstrong World Industries,
    Inc.
Lancaster, PA
(717) 397-0611

(page 56)
Larry Totah
Los Angeles, CA
(213) 467-2927

(page 59)
Lou Goodman
New York, NY
(212) 243-4236

(page 60, bottom)
Peter Shire
Los Angeles, CA
(213) 662-5385

(page 68, bottom)
Sara Olesker
Chicago, IL
(312) 248-9100

INDEX

PHOTOGRAPHY CREDITS